Table of Contents

I0475710

The preamble..02

Prologue..05

Synopsis..08

Imperative Notes....................................12

10 Potions... To reduce 25 lbs, Excess
Weight in a week and 8 Inch Flab on
Tummy Region..19

....

PREAMBLE

Greetings and Congratulation...... You are little away to drop and move down to 25 lbs excess weight in a week and 8 inch waist, the flab on tummy area (indeed, fat will start trimming back in a few days from below the ribs to up hip. Take in a nice and stunning physique form of your name.

Excessive fat anywhere in the torso area and exclusively belly flab remain peculiar and irksome; create a fearful position for a person.

The causes could be junk food taking in carbohydrates in heavy amount or indolent slothful life style. To make out with this nuisance concern, some remain hungry and starved, but tummy remain in the same location with same size and self obsessed treatment make harmful side effects of their physical condition. It can disturb emotional, physical energies and sexual desire as easily....... Sufficient people habitually wish to burn down the flab from belly region and make an incorrect approach to contract out the unnecessary immensity of fat from the tummy department. Those are obese or overweight or having a soft stomach and their timetable simply permits them for deep colors, weighty fast food and soft drinks or similarity of sweet, salty, astringent, or any bitter kind of foodstuff and they want to capture rid of surplus body weight than every potion in this book will positively help them for a sure event.

"Savants"...... World fame holder, provide praiseworthy services to decrease and bearing weight naturally without any side force and their dietary potions are applied for centuries for weight loss and for the slash of belly flab.

In fact, if any one like to scale down and trim down excess body weight and consistent about it than try to complete and thorough assurance without an iota of suspicion about these potions in the book to make rid of flab on tummy area or in a different spot on the physical structure.

Good luck and have a smart, trim shape with healthy physique configuration....

Dr. Mohammad Anees

....

PROLOGUE

Ciao and warm welcome.......... I assure all, those are anxious about losing weight, I did not forget any mark of realistic research lead to weight loss. Every potion in the book to remove excess weight are stumpy in calories, but include vitamins, minerals and pack with fiber.

"In general milieu, weight loss may cause energy fall in the human physical structure... For this reason... A panel of investigators completed organized assessment with meta-analysis of data of more than 1,000 controlled trials to look at the weight loss effects using any portion in the book. Crosswise to the panel studies, their assessment confirmed nearly zero side effects on the health of all these potions..... In addition, panel studies put side by side, these most admired potions for years by "Savants" not just to get rid of the flab from tummy and to slim down fat from different part of the torso but also significantly reduces the risk of diabetes and blood pressure meticulously.

Many fields are already printed in the International Journals to curb and check obesity. Many stay in their flow of studies......... If you delete this during breakfast, lunch and dinner, can escort you to weight loss promptly. Later on, another study generates another endless contradiction and obese people remain in waiting list and anxiously to catch rid of by surplus flesh and flab from their physique configuration. In order to fulfill your purpose.......... Plus enhances your metabolism. Higher metabolism will easily shed the flab from the torso.... As well hold your eyes open with the firm clamp of check to navigate correctly liver function. The liver is the organ that processes fat, and a healthy liver will completely split down the fat that would ultimately store as redundant in cavity anywhere in the physical structure and cause obese.

It's obvious that the precedents behind the above mentioned are for all, those are eager and crave to cut weight.....

Do not waste enormous research further and please follow any potion in the book to definitively end result in weight loss and suffer a happy life with trim/slim build. Almighty God bless all of us.

Dr. Mohammad Anees

....

SYNOPSIS

Since the conception of souls.... As a human existence, all individuals desire to look young with ideal body, like to have wonderful human body structure and its shape.. To obtain marvelous structure and figure... Diet logy remains hush-hush factor of our body beauty. This aspiration is thousands years old. The entire potion for weight loss in this book has passed rigid process with a candid approach of experts by all means through hundreds of years. Beside excellent, exceptional efforts in the arena of weight loss by a group of "Savant" all the potions had influenced the keen and passionate people with such enormous satisfactions that no sickness condition appears baloney. Most of the weight loss would be from wrong advice, instructions / cures / potions, etc.....
Which most likely to hit back when you get off the diet? This isn't a result of weight loss........ Everybody has right to give life

with full comfort. Key solution to take away the flab from body region is to provide an everlasting procedure in good physical shape having zero side effects. Your ration food in each potion and in added notes is included with vitamin, minerals and roughage. Each potion in the script and in the added notes provides nearly 700 - 950 calories per day to keep the rapidity of muscle escalation act and the metabolism's action......

Far away from each other.... Fruity, dry fruit, cutting veggies, protein drinks, chicken and fish, you can eat entirely. Only after you find your desired pattern and keeping in view all the potion in the book with added notes, you can eat afterwards more.

Note: Common foods and drinks such as dairy products, deep brown with cream and tea with milk, regular sodas and soft beverages, white flour breads, potatoes... cooked or baked, not suitable for getting rid of flab or excess weight. Like all sweet dishes in whatever form or shape, are forbidden.

···

Essential:

For lunch and dinner along with each portion of the script.

Get ¼ kg broiler chicken meat from the bosom, and make into two bits. Make steam, roast and keep one piece for lunch with ¼ brown bread (preferable Arabic brown bread) or with a small size plate of rice and another man for dinner.... Repeat the same steps....

Or take tuna fish 100 gram and make in equal parts of 50 grams for lunch and dinner. Put one fresh lemon juice on 50 gram tuna fish every time during lunch, dinner and eat in that instance with little bread not more than ¼ brown bread or small plates of brown rice.

Or obtain a composition of 100 gram trout fish from the grocery store (without skin and thorns) and draw in equal parts of 50 grams. Put one fresh lemon juice on 50 grams of the trout fish piece every time during lunch, dinner and eat in that instance with little bread not more than ¼ brown bread (preferable Arabic brown bread made with whole wheat) or small plates of brown rice.

With all-inclusive kindness to remove flab and excess weight from a body among each portion of the script. Please pay heed to below Notes.

● ●

IF craving with a shocking appetite for more food.

If anyone even feels hungry with a startling craving for more food than take only steamed seafood or boiled broiler chicken not more than 50 grams (A pause, from the breast of Broiler Chicken or in limit of maximum, 50 gram tuna fish).... With ½ brown bread like brown Arabic bread (preferable brown Arabic bread made with whole wheat) or small plates of brown boiled rice. No fry food or additional crispy foodstuff or dry nuts please. As more food will maneuver the weight loss of 25 lbs in a week and 8 inch removal on tummy flab in two weeks. It could demonstrate inconsistency...

....

IMPERATIVE NOTES

Note: 1

Necessary daily intake before breakfast.

Yield 2 (two) tablespoon full olive oil and one tablespoon full black seeds. Mix black seeds in olive oil and leave minimum 6 hours went through with it. In the morning before eating or drinking anything else. Sift black seeds from the oil.

Like a shot get one tablespoon full Acacia honey and mix black seeds in it. Withdraw with 2 glasses of warm mineral water or if you are living near the sea shore and your water supply is from saline water conversion authority than take and drink 2 glasses of saline purified drinkable warm water.

Get hold of pure extract of aniseed (Fennel seed extract) and drink two tablespoons full with additional small glass (100 g) of water after above course

Note: 2

Throughout the day.

Get 100 gram Apricot

50 grams Prunes

100 g Papaya....

100 g Apple

100 g Peach

100 g Guava

100 g Pear

100 g, Fig

100 g Strawberry

Black pepper powder 15 g

Cinnamon powder 15 g

Preparation:

Peel 100 gram papaya, apples, pears and put in the bowl after slicing..... Cut guava and take out the seeds and slice it...... Also slice into pieces, apricot, prunes, fig, peach and strawberry. Mix all together thoroughly. Sprinkle black pepper powder and cinnamon powder on it thoroughly. Eat during the daylight, when feel alarming hungry and craves for food and drink below stated mixture as you asked. To prepare this mixture to drink......... put all the below mentioned vinegars in 3 liter water..... Stir well and break up properly before drinking.

Apple Cider Vinegar 50 g

Red grapes Vinegar 50 g

White Vinegar (made with cucumber) 50 g

Rice Vinegar 50 g

White grapes Vinegar 50 g

Water 3 Liters

Preparation:

Mix above five vinegars in 3 liter water, liquefy completely and drink throughout the daytime

....

Note: 3

After Dinner and before going to bed

Take one little glass of goat milk. (Not more than ¼ liter)

Take little goat milk from glass and put (one) gram saffron in it. Mix saffron completely by grinding with spoon, holding spoon in hand until the milk becomes colored. Put this with other milk in the glass and drink it without adding anything else. One hour, after taking milk........ Get hold of pure extract of aniseed (Fennel seed extract) 2 (two) tablespoon full without adding any additional detail. (No water please).... And start to kip.

....

Note: 4

A mixture of spices / Condiments / substances for sprinkling daily on items in every potion.

Turmeric powder 20 g

Black pepper powder 20 g

Cinnamon powder 20 g

Dry ginger 20 g

 Beatle leaf nut 10 g

Chinese Chives (Garlic cloves) 20 g

Star Anise powder 10

Preparation:

Take Beatle leaf nut and toast in pan on fire without oil. After crush and grind it.

Mix all above condiments in powder form with ground Beatle leaf nut.

....

Essential note: Don't get overwhelmed or confused...... Strictly refrain from Fried chicken, fish, meat or other kind of fries please.....

....

Potion 1:

Items needed daily to reduce 25 lbs per week and 8 inch waist in two weeks.

Grape leafs 150 g

Watercress 150 g

Grapefruit rind 100 grams

Lemon 50 gram

Marinade 25 g

African Tea   Only 2 cups

Preparation:

Dry grapefruit rind under the sunshine and after grinding all rinds... keep this powder in a bottle.

In the morning after taking a necessary drink as given in Note 1..... Following eating fruit, drinking liquefies water with vinegars, as mentioned in Note 2

Cut watercress in pieces. Take a pot and put 50 g water and 25 g marinade and dissolve it. Put watercress in marinade water for 6 hours..... (A marinade must be of soy and lime).

Boil grape leaves and turn off into slices. Bring out the watercress and mix with grape leaves. Pour 50 gram lemon on it. Sprinkle all the powder mixture of dry grapefruit rinds from bottle on it. As well put all condiments / substances on it as given in Note 4. Eat half at lunch and half at dinner. You can cook fresh for lunch and dinner with half quantity of all particulars.

Pursue Note 1, Note 2, and Note 3 strictly please.

Drink 1 cup African tea every time, after lunch and dinner if you desire without sugar and other sweet elements.

....

Potion 2:

Items needed daily to reduce 25 lbs per week and 8 inch waist in two weeks.

Peaches 100 g

Turnip 150 g

Fennel Vegetable 100 g (Bulb shape plant grows in the ground.... It's different than fennel seeds)

Mustard Paste 50 g

Tomato 200 gram

White vinegar of cucumber 50 g

Lemon 50 gram

Preparation:

Take peaches, cut it in slices, remove the pit and place below the sunshine to complete dryness. After drying up, dig it in powder form and save in a bottle or in plate for use.

Take white cucumber vinegar 50 g and mix with 50 g water.

Peel off the turnip and cut into pieces with fennel bulbs and put in the mixed water for the whole night.

In the morning after taking a necessary drink as given in Note 1..... Following eating fruit, drinking liquefies water with vinegars, as mentioned in Note 2

Smash tomatoes. Put sliced turnip and cook slightly with smash tomatoes. Mix with sliced fennel vegetable. Pour 50 gram lemon on it with ground peaches on it.

Add mustard paste in it. Sprinkle all the condiments / substances on it as given in Note 4. Eat half at lunch and half at dinner with ½ brown bread (preferable whole wheat dough) or with small plate of brown rice. You can cook fresh for lunch and for dinner with half quantity of all particulars.

Pursue Note 1, Note 2, and Note 3 strictly please.

....

Potion 3:

Items needed daily to reduce 25 lbs per week and 8 inch waist in two weeks.

Carrots 150 g

Bottle gourd 150 g

Ridge gourd 150

Turmeric Fresh Vegetable 150 g

Broad or Faya bean 100 g

Tomato 200 g

Marinade (lime) 50 g

Preparation:

Take carrots, grate all and put under the sun to complete dryness. After drying up, break down the carrots and make in powder form.

Peel away the bottle gourd, ridge gourd and fresh turmeric vegetable and cut into slices. Place 50 g Marinade (lime) on it and leave for 2 hours.

Boil broad or Faya bean.

In the morning after removing a necessary drink as given in Note 1..... Following eating fruit, drinking liquefies water with vinegars, as mentioned in Note 2

Smash tomatoes. Put Sliced bottle gourd, ridge gourd and turmeric vegetable and cook slightly with smash tomatoes. Mix with boiled broad beans. Pour 50 gram lemon on it. Add powdered carrots and sprinkle all the condiments / substances on it as given in Note 4. Eat half at lunch and half at dinner with ½ brown bread (preferable whole wheat bread) or small plate of rice. As well you can cook fresh for lunch and for dinner with half quantity of all particulars.

Pursue Note 1, Note 2, and Note 3 strictly please.

....

Potion 4:

Items needed daily to reduce 25 lbs per week and 8 inch waist in two weeks.

Bitter gourd 250 g

Amaranth Vegetable 150 g (It's like green peas, but has green beans with the black eye)

Tomato 250 gram

Lime Marinade 50 gram

Smoked sea salt 10 g

White vinegar (Rice) 20 GM

Preparation:

Clean top layer of bitter gourd and divides in two pieces lengthwise. Contract away the germs and keep it separate from cooking. Put lime Marinade on it with 10 g smoke sea salt and leave for 4 hours.

Sliced fresh amaranth vegetable and keep it separate.

Take 20 g white vinegar on Rice and put in 100 g water. After mixing, saturate sliced amaranth vegetable in it also for 4 hours.

In the morning after taking a necessary drink as given in Note 1..... Following eating fruit, drinking liquefies water with vinegars, as mentioned in Note 2

Smash tomatoes. Put sliced bitter gourd with seeds and remaining water of the lime marinade in the pot (Bitter gourd beans must be cleaned thoroughly) and cooked in smash tomatoes. Mix with sliced fresh amaranth vegetable. (Please drink, if whatever mixture with vinegar remain) Pour 50 gram lemon on it. Sprinkle all the condiments / substances on it as given in Note 4. Eat half at lunch and half at dinner with ½ brown bread (preferable whole wheat dough) or small plate of rice. As well you can cook fresh for lunch and for dinner with half quantity of all particulars.

Pursue Note 1, Note 2, and Note 3 strictly please.

....

Potion 5:

Items needed daily to reduce 25 lbs per week and 8 inch waist in two weeks.

Okra 100 g

Lotus stems 200 g

Black eye white beans 150 g

Tomato 250 gram

Lime Pepper 20 gram

Smoked sea salt 10 g

Mustard paste 20 grams

Peach 50 g

Preparation:

Take all okras, remove cover and underside.... Turn off into slices and place under the sunshine for complete dry. After drying up, grind all dried okras in powder form and keep separate in a dish or bottle.

Clean top layer of lotus root and divides in small shards. Place 10 g smokes sea salt and leave for 2 hours.

Boil black eye beans and strain it, shake off the water aside.

Take peach and smash it. Mix with mustard paste..... Makes sauce and keep in the separate bowl.

In the morning after removing a necessary drink as given in Note 1..... Following eating fruit, drinking liquefies water with vinegars, as mentioned in Note 2

Put divided lotus stem and cook in smashed 250 g tomatoes. Put two spoonful olive oil and mix boiled black eyed beans in it after lotus stem become tender. Sprinkle 20 gram lime pepper and ground okras powder with all the condiments / substances on it as given in Note 4. Take peach /mustard sauce and lotus stem / black eye bean dish and eat half at lunch and half at dinner with ½ brown bread (preferable whole wheat bread) or a modest plate of rice. As well you can cook fresh for lunch and for dinner with half quantity of all particulars.

Pursue Note 1, Note 2, and Note 3 strictly please.

....

Potion 6:

Items needed daily to reduce 25 lbs per week and 8 inch waist in two weeks.

Aloe Vera 100 g

Chicken from breast without ribs 100 g

Green peas 100 g

Marcella Tomato Sauce 150 gram

Lime Pepper 20 gram

Smoked sea salt 10 g

Dry fenugreek 20 g

Chinese Chives (garlic cloves) 50 gram

Preparation:

Contain Aloe Vera, remove the peel, keep only kernel and hit it.

Fresh boneless chicken from breast and slice into little slices.

Put water ¼ liter in a stack with 50 g chicken pieces in it with green peas and Aloe Vera. Add smoke sea salt and lime pepper, garlic cloves too. Boil all till water remain half.

In the morning after taking a necessary drink as given in Note 1..... Following eating fruit, drinking liquefies water with vinegars, as mentioned in Note 2

Steaming roast other 50 gram chicken and eat half piece 25 g with Marcella tomato sauce with ½ brown bread (preferable whole wheat bread)) or a modest plate of rice.. In addition, with ½ quantities of chicken and green peas, Aloe Vera soup for lunch from pot...... In soup..... Please add a full measure of dry fenugreek with all the condiment / substances as given in Note 4. Heat and stir until it thickens. You can cook fresh for lunch and for dinner with half quantity of all particulars.

Pursue Note 1, Note 2, and Note 3 strictly please.

....

Potion 7:

Items needed daily to reduce 25 lbs per week and 8 inch waist in two weeks.

Green lintel 200 g

Tuna fish 100 g (precooked)

Dried Lemon 50 g

Mint 50 g

Lime Pepper 20 gram

Smoked sea salt 10 g

Ginger 100 g

Chinese Chives (garlic cloves) 50 GM

Preparation:

Clean, green lentil and soak in water for 5 hours.

Skin the ginger pieces and garlic cloves, Smash, make a paste and keep in small dish.

Put water ½ liter water in a raft with 200 g lintel Add smoked sea salt, lime pepper, garlic cloves and ginger pieces paste. Cook all lintel until tendered and water remain little.

In the morning after taking a necessary drink as given in Note 1..... Following eating fruit, drinking liquefies water with vinegars, as mentioned in Note 2

Consume 100 g tuna fish on plate and put 50 g chopped dried lemon with 50 g mint on it. Take lintel out from the beach and lay all the lintel in the arena. Add all the condiments / substances in it as given in Note 4. Eat ½ Tuna dish and ½ quantity of lintel with brown bread (preferable whole wheat dough) or small plate of rice. Keep remaining quantity for dinner. As well you can cook fresh for lunch and for dinner with half quantity of all particulars.

Pursue Note 1, Note 2, and Note 3 strictly please.

....

Potion 8:

Items needed daily to reduce 25 lbs per week and 8 inch waist in two weeks.

Bran (Husk from wheat, you can sift brown flour by yourself to get bran) 200 g

Black eye beans 100 g

Whole fennel seeds 50 g

Mint 50 g

Smoked sea salt 10 g

Ginger 100 g

Chinese Chives (garlic cloves) 50 gram

Preparation:

Boil black eye beans and smash completely into a paste...

Take whole fennel seeds and grind all in powder form.

Mix both with bran and mold in 6 equal biscuits for baking. Keep separate 6 biscuits of equal size after baking in covered dish.

Peel the ginger pieces and garlic cloves with mint, Smash and make sauce. Keep separate in small dish.

In the morning after taking a necessary drink as given in Note 1..... Following eating fruit, drinking liquefies water with vinegars, as mentioned in Note 2

Remove garlic, cloves and mint sauce and add all the condiment / substances in it as given in Note 4.... Eat ½ ginger / garlic sauce with 3 biscuits at lunch and ½ sauce with 3 biscuits at dinner.

Pursue Note 1, Note 2, and Note 3 strictly please.

....

Potion 9:

Items needed daily to reduce 25 lbs per week and 8 inch waist in two weeks.

Carrot 100 g

Black berry leafs 100 g

Cucumber 100 g

Dried Whole French Beans 100 g.

Ridge gourd 100 g

Smoked sea salt 10 g

Ginger 100 g

Chinese Chives (garlic cloves) 50 GM

Preparation:

Boil blackberry leaves and ridge gourd together. After boiling, drop the water aside and both vegetable cut into slices.

Peel ginger and garlic cloves, dash into a paste. Add smoked sea salt in it and blend with vegetable pieces and keep in a bowl or flowerpot.

Get hold of dried whole French beans, toast in pan on fire without oil and grind in powder form.

In the morning after taking a necessary drink as given in Note 1..... Following eating fruit, drinking liquefies water with vinegars, as mentioned in Note 2

Remove carrots and cucumbers. Cut into pieces. Add ground dried whole French beans in it with all the condiments / substances in it as given in Note 4...

Mix vegetable dish and carrot, cucumber dish together. Eat half at lunch and half at dinner with ½ brown bread (preferable whole wheat bread) or small plate of rice. As well you can cook fresh for lunch and for dinner with half quantity of all particulars.

Pursue Note 1, Note 2, and Note 3 strictly please.

....

Potion 10:

Items needed daily to reduce 25 lbs per week and 8 inch waist in two weeks.

Blue berry leafs 50 g

Can berry leafs 50 g

Bar berry leafs 50 g

Bear berry leafs 50 g

Mel berry leafs 50 g

Crow berry leafs 50 g

Apple wood smoke sea salt 10 g

Smoke marinates 20 g

Ginger 50 g

Chinese Chives (garlic cloves) 50 GM

Dried lemon 50 g

Preparation:

Take dried lemon and crush into powder form.

Boil all berry leaves and turn off into slices. Put smoke marinates on it and keep in separate bowl.

Peel ginger and garlic cloves, dash into a paste. Add apple wood smoke sea salt in it and mix with berry leafs pieces in the bowl or flowerpot.

In the morning after taking a necessary drink as given in Note 1..... Following eating fruit, drinking liquefies water with vinegars, as mentioned in Note 2...

Take berry leafs bowl and add all the condiment / substances in it as consecrated in

Note 4...

Mix dried lemon powder in berry leaves. Eat half at lunch and half at dinner with ½ brown bread (preferable whole wheat dough) or small plate of rice. As well you can cook fresh for lunch and for dinner with half quantity of all particulars.

Pursue Note 1, Note 2, and Note 3 strictly please.

....

End Note

Splendor thanks for picking out this book to reduce 25 lbs extra weight in a week and 8 inch belly flab in two weeks (exclusively from ribs to up buttocks)......... Regularity is needed to accomplish your objective of body condition and reduction in excess weight.

Continue follow any potion in the book to achieve the object of reducing weight 25 lbs in a week and 8 inch flab from your bay window section.

Almighty God bless all of us with wonderful health and ideal physique appearance.

Dr. Mohammad Anees

www.ingramcontent.com/pod-product-compliance
Lightning Source LLC
Chambersburg PA
CBHW021940170526
45157CB00005B/2369